BACK FROM THE BRINK

SEA ANIMALS

BLACK RABBIT BOOKS

JOANNE MATTERN

TABLE OF CONTENTS

1

Bluefin Tuna

4

This big fish once swam all over the Atlantic Ocean. But overfishing cut down its numbers. Something had to be done.

The government limited how many fish someone could catch. Scientists **tagged** the fish. They learned where the fish lived. This helped them protect the fish's home. As of 2021, bluefin tuna are no longer at risk.

Did You Know?

Bluefin tuna grow to be about 6 feet (1.8 meters) long.

2

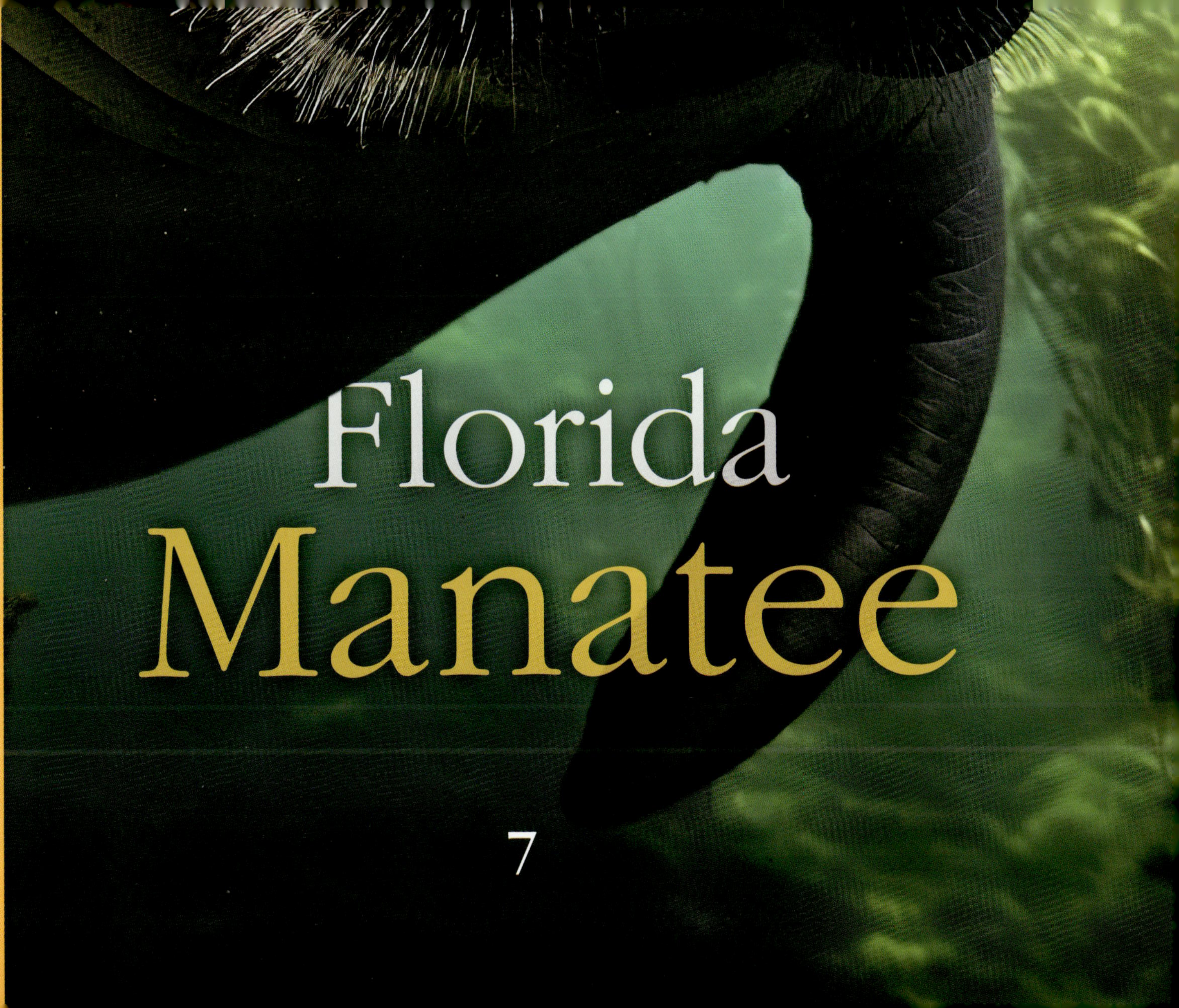
Florida
Manatee

Manatees swim in warm waters in the southern United States. People also like these waters. They drive boats near the manatees' homes. Many animals are hit and killed.

There are laws to protect the manatees. Some places have slow speed spots. People slow their boats' speed. They look for manatees and go around them.

The manatee is still in trouble. But this sea creature is coming back because of safer waters.

Think About It

Why are laws important for protecting animals?

3 Guadalupe Fur Seal

This seal lives in the Pacific Ocean. Long ago, many seals were killed for their fur. People thought the animals were gone. Then in 1954, scientists found some seals in a cave.

Mexico and the United States rushed to save the seal. They made it **illegal** to hunt the animal. The number of fur seals grew. Today, these animals are doing well.

Did You Know?
The fur seal can dive to a depth of 180 feet (55 m).

4

Humpback Whale

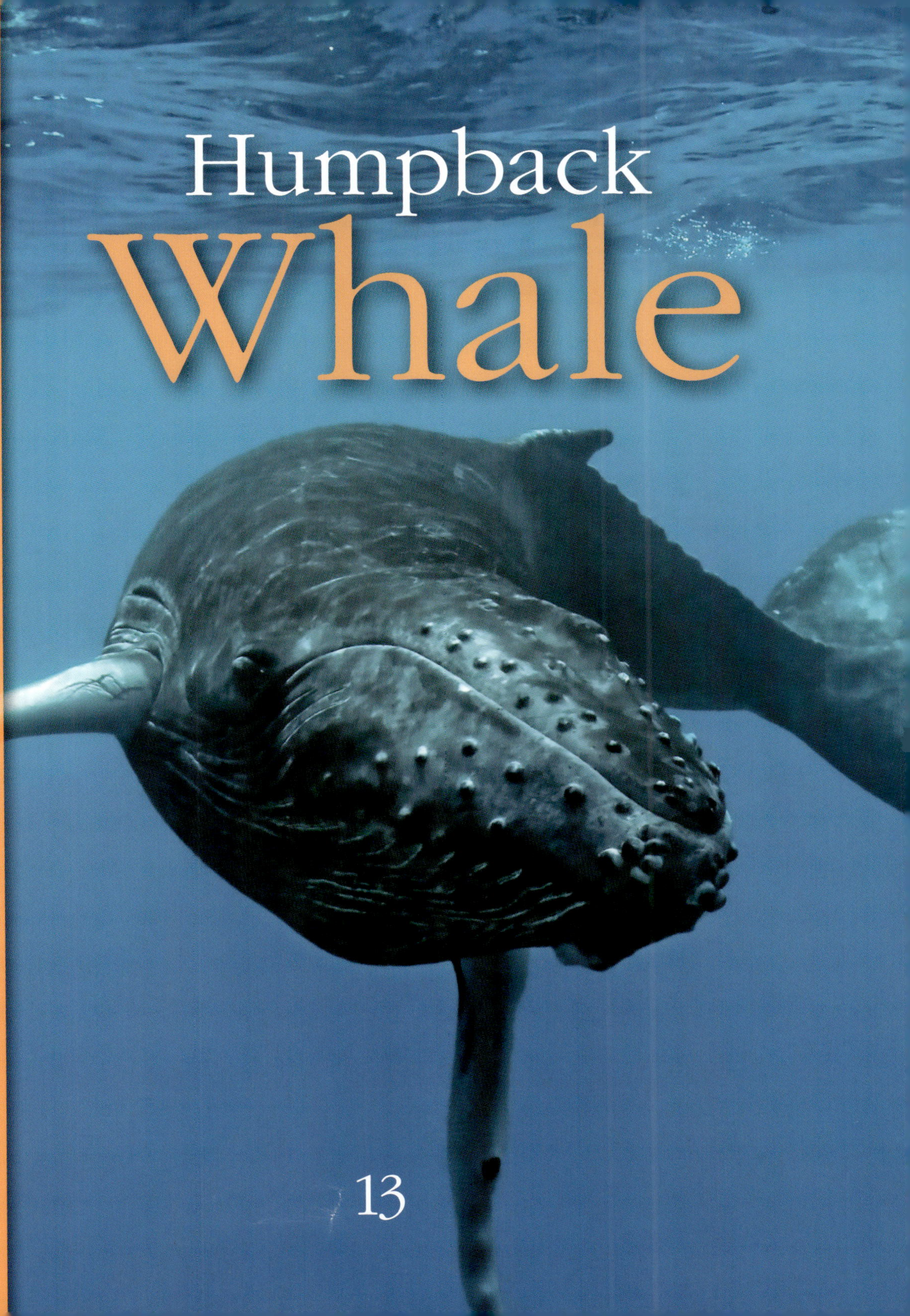

For years, these giant whales were hunted for their meat and blubber. By the 1980s, there were just 440 humpbacks left.

Many countries began to ban hunting whales in 1986. This was great news for the humpback. Today, large numbers of these whales swim through the world's oceans.

Did You Know?
The Endangered Species Act is a list of animals that are protected by US law. Humpbacks were added in 1973.

Steller Sea Lion

5

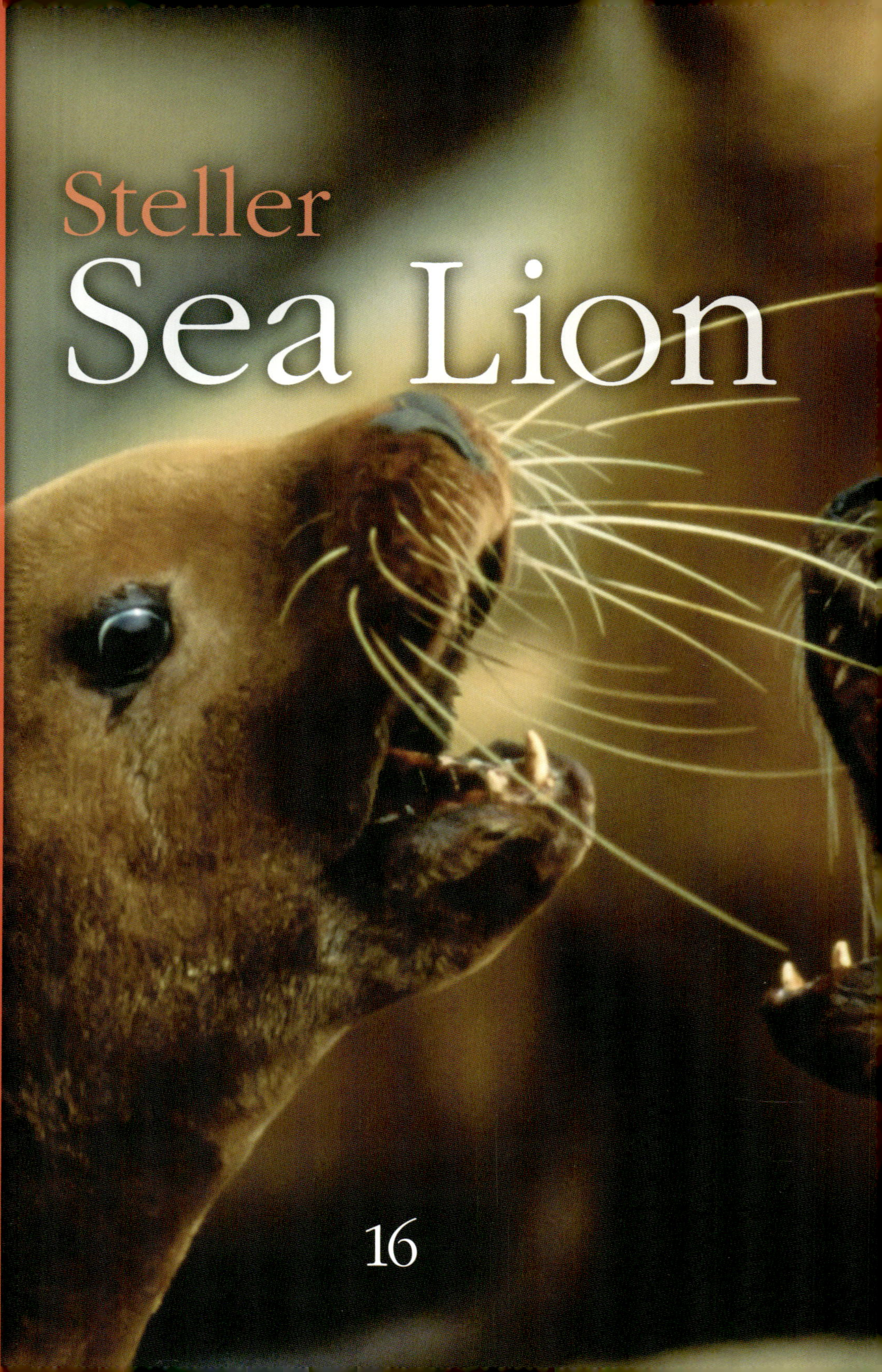

These animals live in the cold waters of the northern Pacific Ocean. There are two types of sea lions. The eastern sea lions are doing well. But the western sea lions were dying out.

Overfishing caused sea lions to lose their **prey**. So the US government passed laws. They made hunting the sea lions illegal. They also put a limit on how many fish people could catch. Today, sea lions are doing much better.

Think About It
How can the loss of prey affect animals?

6

Swordfish

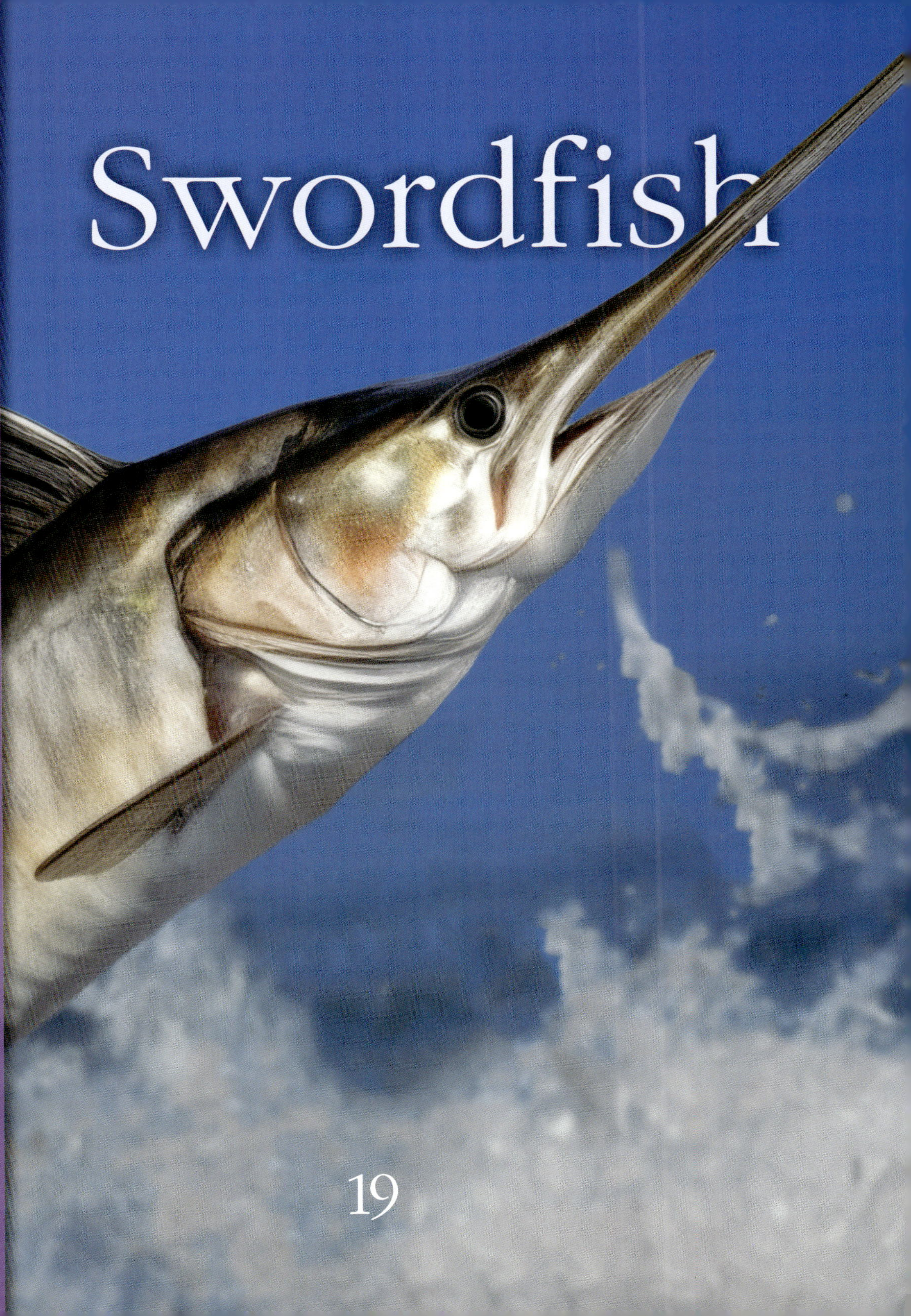

The swordfish gets its name from its long beak. It is a top predator. But it is also a top target for fishing. By 1999, scientists worried that the fish would die out.

Countries around the world worked to save the fish. They limited how many swordfish could be caught. Scientists also use satellites to track swordfish. This helps them keep count of the fish. By 2017, there were a lot more swordfish in the oceans.

Think About It
What other ways can technology help at-risk animals?

MORE TO EXPLORE

WHERE IN THE WORLD?

GUADALUPE FUR SEAL
Pacific Ocean off the west coast of North America

FLORIDA MANATEE
Coast of Florida and southern United States

BLUEFIN TUNA
Atlantic, Pacific, and Indian Oceans

SWORDFISH
Atlantic, Pacific, and Indian Oceans

HUMPBACK WHALE
All oceans

STELLER SEA LION
Along the coasts of the northern Pacific Ocean

MORE TO EXPLORE

FANTASTIC FACTS

The bluefin tuna is one of the heaviest fish in the ocean.

Manatees are also called sea cows.

Guadalupe fur seals dive deep to feed on fish and squid.

Whale blubber was once used to make oil for lamps.

Steller sea lions can climb cliffs and rocks.

Swordfish can swim as fast as 60 miles (97 kilometers) per hour.

MORE TO EXPLORE

COOL COMPARISONS

These animals are making a comeback. How many live in the wild today?

Bluefin Tuna
more than 1 million

Florida Manatee
between 7,000 and 11,000

Guadalupe Fur Seal
about 34,000

Humpback Whale
about 84,000

Steller Sea Lion
about 46,000

Swordfish
unknown, but numbers are rising

MORE TO EXPLORE

RESOURCES

Glossary

blubber (BLUB-uhr) The fat of whales and other large sea mammals.

endangered (en-DAYN-jurd) Close to no longer existing.

illegal (ih-LEE-guhl) Something that is against the law or rules.

predator (PRED-uh-tuhr) An animal that hunts and eats other animals.

prey (PRAY) An animal hunted or killed for food.

satellite (SA-tuh-layht) A machine sent into space that moves around the Earth, moon, sun, or another planet.

tag (TAG) To attach a plastic device on an animal so it can be identified.

Read More

Dufresne, Emilie. *Endangered Animals in the Sea.* New York: PowerKids Press, 2021.

Gleisner, Jenna Lee. *Humpback Whales.* Minneapolis: Jump!, Inc., 2024.

Index

TOP RANK is published by Black Rabbit Books, P.O. Box 227, Mankato, MN, 56002. •

• Top Rank is an imprint of Black Rabbit Books • Edited by Ana Brauer • Series designed by Danny Nanos • Book designed by Jason Knudson • Photographs © Adobe Stock/Yendi08, 19, 21, 23; Alamy Stock Photo/Alex Mustard, 5, Joel Sartore Photography, 16–17, Michael Melford, 18, 21, 23, WaterFrame_fba 20, 21, 23; Dreamstime/Bualong Sadsana, 6-7, 21, 23; Getty Images/ Michael Nolan/ robertharding, 12, 21, 23; Minden Pictures/Brian Skerry, 4, 21, 23, Nick Garbutt / NPL, 14; Shutterstock/Andrea Izzotti, 8, boulham, cover, 1, Justin Berken, 15, Stock High angle view, 2, Sunshower Shots, 9, Wirestock Creators, cover, 1; SuperStock/BlueGreen Pictures, 10-11, Norbert Probst, 13, 21, 23 • Printed in India

Library of Congress Cataloging-in-Publication Data: Names: Mattern, Joanne, 1963- author. | Title: Sea animals: back from the brink / by Joanne Mattern. | Description: Mankato, MN: Top Rank is an imprint of Black Rabbit Books, [2026] | Series: Back from the brink | Includes bibliographical references and index. | Audience: Ages 8-11 | Audience: Grades 2-3 | Identifiers: LCCN 2024043393 (print) | LCCN 2024043394 (ebook) | ISBN 9781644668016 (library binding) | ISBN 9781644668337 (paperback) | ISBN 9781644668658 (ebook) | Subjects: LCSH: Marine animals—Juvenile literature. | Marine animals—Conservation—Juvenile literature. | Endangered species—Juvenile literature. | Classification: LCC QH541.5.S3 M335 2026 (print) | LCC QH541.5.S3 (ebook) | DDC 591.68—dc23/eng/20250110 | LC record available at https://lccn.loc.gov/2024043393 | LC ebook record available at https://lccn.loc.gov/2024043394